AF575500

Backyard Birds
Crows
by Rachael Barnes
Blastoff! Readers
1
Bellwether Media • Minneapolis, MN

Blastoff! Readers are carefully developed by literacy experts to build reading stamina and move students toward fluency by combining standards-based content with developmentally appropriate text.

Level 1 provides the most support through repetition of high-frequency words, light text, predictable sentence patterns, and strong visual support.

Level 2 offers early readers a bit more challenge through varied sentences, increased text load, and text-supportive special features.

Level 3 advances early-fluent readers toward fluency through increased text load, less reliance on photos, advancing concepts, longer sentences, and more complex special features.

★ **Blastoff! Universe**

Reading Level

Blastoff! Beginners — Grade K

Blastoff! Readers — Grades 1–3

Blastoff! Discovery — Grade 4

This edition first published in 2023 by Bellwether Media, Inc.

Library of Congress Cataloging-in-Publication Data

Names: Barnes, Rachael, author.
Title: Crows / Rachael Barnes.
Description: Minneapolis, MN : Bellwether Media, 2023. | Series: Backyard birds | Includes bibliographical references and index. | Audience: Ages 5-8 | Audience: Grades K-1 | Summary: "Developed by literacy experts for students in kindergarten through grade three, this book introduces crows to young readers through leveled text and related photos"– Provided by publisher.
Identifiers: LCCN 2022002362 (print) | LCCN 2022002363 (ebook) | ISBN 9781644876916 (library binding) | ISBN 9781648347375 (ebook)
Subjects: LCSH: Crows–Juvenile literature.
Classification: LCC QL696.P2367 B29 2023 (print) | LCC QL696.P2367 (ebook) | DDC 598.8/64–dc23/eng/20220125
LC record available at https://lccn.loc.gov/2022002362
LC ebook record available at https://lccn.loc.gov/2022002363

Editor: Rebecca Sabelko   Designer: Laura Sowers

Printed in the United States of America, North Mankato, MN.

Table of Contents

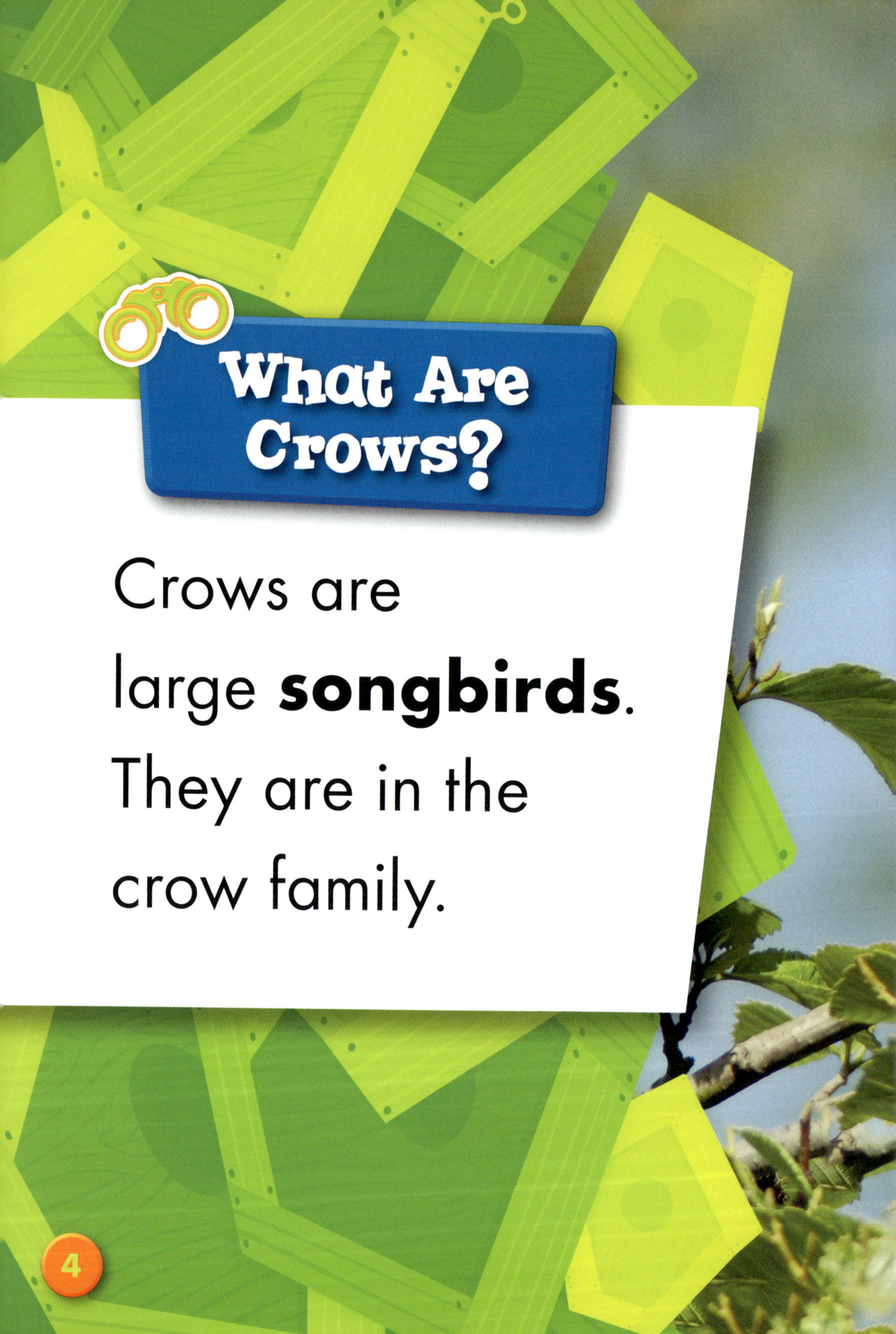

What Are Crows?

Crows are large **songbirds**. They are in the crow family.

All in the Family
blue jay
yellow-billed magpie
common raven

Crows are all black. Their feathers are shiny. Their legs are long.

All Over the Place

Crows live near forests and cities. They build nests high up in trees.

nest

Crows walk on the ground to find food.

Crows eat fruit and **insects**. They sometimes steal eggs and dog food!

Crow Food
fruit
insects
eggs

Feathered Families

Crows live in family groups. They find food and stay safe together.

In winter, crows form **roosts**. They keep each other warm.

roost

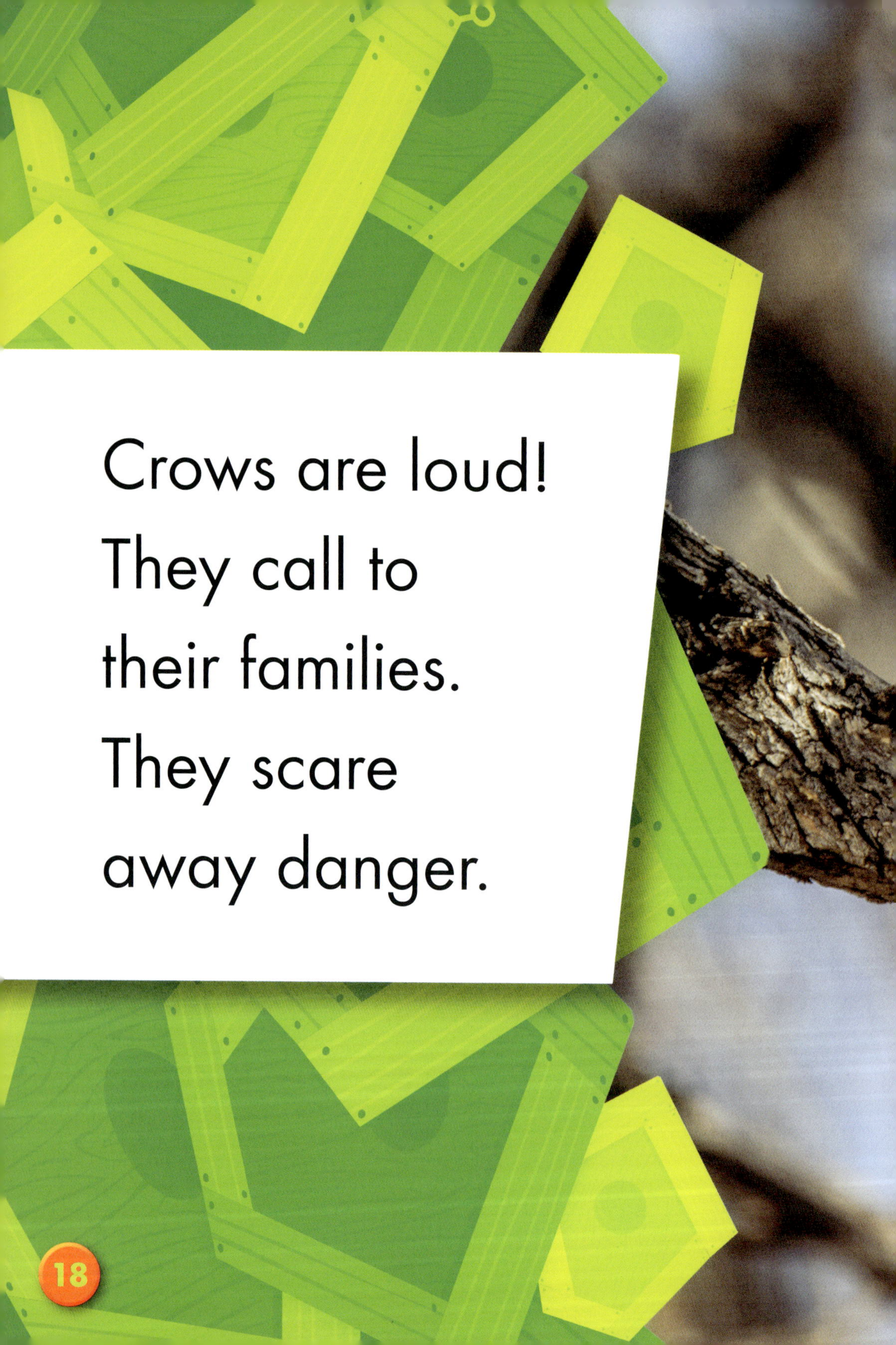

Crows are loud! They call to their families. They scare away danger.

Crow Call
caw!
caw!
caw!

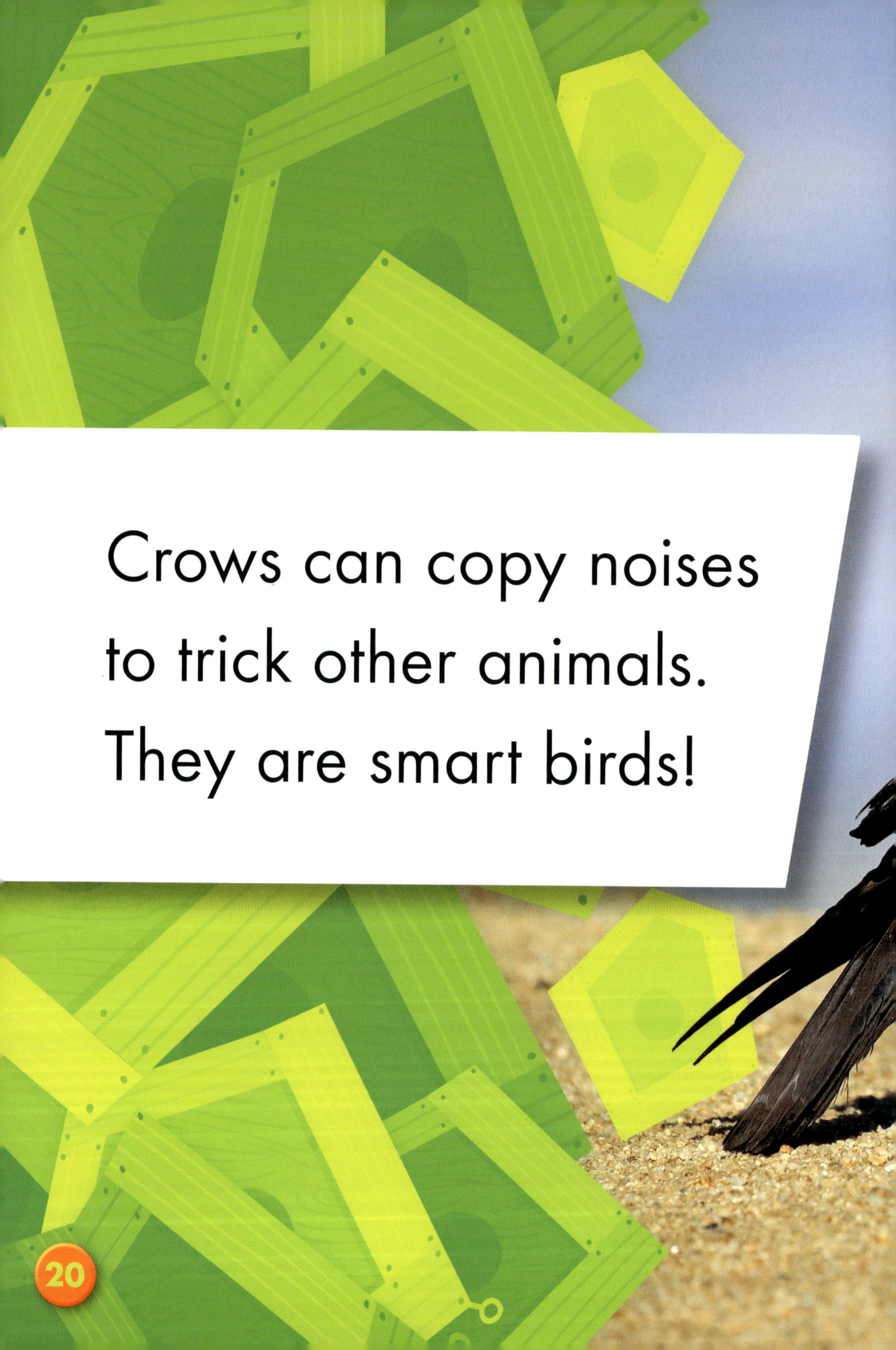

Crows can copy noises to trick other animals. They are smart birds!

Glossary

insects

small animals with six legs and hard outer bodies

songbirds

birds that make musical sounds

roosts

groups of birds that rest together; a crow roost can have thousands of birds.

To Learn More

AT THE LIBRARY

Neuenfeldt, Elizabeth. *Blue Jays.* Minneapolis, Minn.: Bellwether Media, 2022.

Savage, Candace. *Hello, Crow!* Vancouver, B.C.: Greystone Kids, 2019.

Ward, Jennifer. *How to Find a Bird.* New York, N.Y.: Beach Lane Books, 2020.

ON THE WEB

FACTSURFER

Factsurfer.com gives you a safe, fun way to find more information.

1. Go to www.factsurfer.com.
2. Enter "crows" into the search box and click 🔍.
3. Select your book cover to see a list of related content.

Index

The images in this book are reproduced through the courtesy of: BGSmith, front cover (crow); Nanisimova, front cover (backyard); SF photo, p. 3; Keith Lejon, pp. 4-5; FotoRequest, p. 5 (blue jay); Hayley Crews, p. 5 (yellow-billed magpie); Piotr Krzeslak, p. 5 (common raven); Melinda Fawver, pp. 6-7; Richard Mittleman/ Gon2Foto/ Alamy, pp. 8-9; Kresimir Durlen/ Alamy, p. 9 (nest); Susan Hodgson, pp. 10-11; Brian Kushner/ Alamy, pp. 12-13; Adish Zainer, p. 13 (fruit); Joseph Skompski, p. 13 (insects); Bruce MacQueen, p. 13 (eggs); Karen Hogan, pp. 14-15; Bob Gibbons/ Alamy, pp. 16-17; Gerald A. DeBoer, pp. 18-19; imageBroker/ Alamy, pp. 20-21; Silvia Dubois, p. 22 (insects); Elliotte Rusty Harold, p. 22 (roosts); Chase D'animulls, p. 22 (songbirds); Arend Trent, p. 23.